Meet a Baby Polar Bear

Whitney Sanderson

Lerner Publications • Minneapolis

Lerner Publications Company
An imprint of Lerner Publishing Group, Inc.
241 First Avenue North
Minneapolis, MN 55401 USA

For reading levels and more information, look up this title at www.lernerbooks.com.

Main body text set in Billy Infant Regular.
Typeface provided by SparkType.

Map illustration on page 20 by Laura K. Westlund.

Library of Congress Cataloging-in-Publication Data

Names: Sanderson, Whitney, author.
Title: Meet a baby polar bear / Whitney Sanderson.
Description: Minneapolis : Lerner Publications, [2024] | Series: Lightning bolt books—baby North American animals | Includes bibliographical references and index. | Audience: Ages 6-9 | Audience: Grades 2-3 | Summary: "How big is a baby polar bear? How does it change as it grows? Readers will learn these things and more in this book full of fun facts and cute-as-can-be photos of baby polar bears"— Provided by publisher.
Identifiers: LCCN 2022034642 (print) | LCCN 2022034643 (ebook) | ISBN 9781728491103 (library binding) | ISBN 9781728498461 (ebook)
Subjects: LCSH: Polar bear—Infancy—North America—Juvenile literature.
Classification: LCC QL737.C27 S253 2024 (print) | LCC QL737.C27 (ebook) | DDC 599.78613/92—dc23/eng/20220721

LC record available at https://lccn.loc.gov/2022034642
LC ebook record available at https://lccn.loc.gov/2022034643

Manufactured in the United States of America
1-53041-51059-11/14/2022

Table of Contents

Baby Bear

A polar bear nestles in her den in the snow. The heat of her body keeps her babies warm. Baby polar bears are called cubs. They grow inside their mother for about eight months.

Newborn cubs weigh about 1 pound (0.5 kg). They are as big as a jar of peanut butter. Short, soft fur covers their bodies.

Cubs drink milk from their mother. Polar bear milk is 30 percent fat. Cow's milk is only 3 to 4 percent fat. The rich milk helps the cubs stay warm.

Mother polar bears take care of their cubs. Fathers do not help raise them. Two to three months after they are born, cubs leave the den. They are about the size of a large watermelon.

Cubs stay near their mother after leaving the den.

Mom and Baby

Polar bears live in the Arctic Circle. Temperatures get as low as -50°F (-46°C). That is much colder than a freezer!

Cubs learn to walk in the snow. They wrestle and chase one another across the tundra. Sometimes their mother carries them in her mouth or on her back.

Polar bears communicate with sounds. They grunt, growl, roar, purr, and chuff.

No animals hunt adult polar bears. But wolves or other polar bears might attack a cub. Their mother has a good sense of hearing, smell, and sight. She teaches her cubs to avoid danger.

Cubs also learn to swim.
A polar bear's large,
webbed feet help it swim
for long distances.

Ice Hunter

Every fall, ice forms on the Arctic Ocean. Polar bears hunt on sea ice. When cubs are three to four months old, they head onto the ice with their mother.

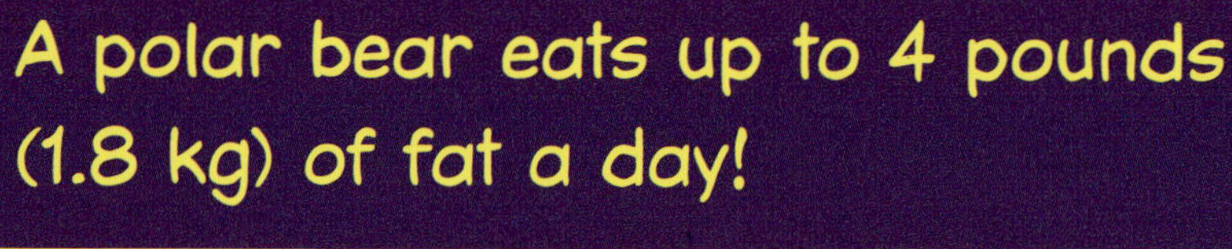

Cubs keep drinking milk until they are about two years old. They also learn to hunt. They mainly eat seals. They will also eat berries, seaweed, or meat left by other predators.

Cubs learn to hunt by watching their mother. She has two ways of hunting. She might sneak up on a seal resting on the ice. Or she might wait at the edge of the ice for a seal to come up for air. Then she attacks!

Arctic Explorer

A baby polar bear stays with its mom for two and a half years. Then it can catch its own food.

A young male polar bear can grow to be 1,300 pounds (590 kg). That's about the size of a dairy cow. Female polar bears can grow to be 650 pounds (295 kg).

Polar bears are tough.

Their large size, sharp teeth, and claws help them catch prey. Two layers of fur keep them warm in the Arctic.

Polar bears can live for fifteen years or more.

Adult polar bears hunt and find mates. They have babies when they are about five years old.

Polar Bear Life Cycle

Habitat in Focus

- In North America, polar bears live in Canada and Alaska.

- Climate change is causing more sea ice to melt each year. Polar bears have less food and territory. Scientists are working to protect the sea ice.

- Polar bears don't need to drink water! They get water from digesting the fat they eat.

Fun Facts

- A polar bear standing on its hind legs is up to 11 feet (3.3 m) tall.

- Polar bears are the biggest land predator in the world.

- Polar bear fur looks white, but it is transparent, or clear.

- How far can a polar bear swim? The record is 426 miles (686 km).

Glossary

chuff: to breathe with a puffing sound

communicate: to use sounds or actions to share a message

den: the resting or living place of a wild animal

predator: an animal that hunts other animals for food

prey: an animal that another animal hunts for food

tundra: flat, frozen land with no trees

webbed: connected by skin

Learn More

Ardely, Anthony. *What If Polar Bears Disappeared?* New York: Gareth Stevens, 2023.

Hansen, Grace. *Help the Polar Bears.* Minneapolis: Abdo Kids Jumbo, 2019.

Kiddle: Polar Bear Facts for Kids
https://kids.kiddle.co/Polar_bear

Murray, Tamika M. *Meet a Baby Gray Wolf.* Minneapolis: Lerner Publications, 2024.

National Geographic Kids: Polar Bear
https://kids.nationalgeographic.com/animals/mammals/facts/polar-bear

Polar Bears International: Polar Bear Tracker
https://polarbearsinternational.org/polar-bear-tracker

Index

Photo Acknowledgments

Image credits: Art Wolfe/Getty Images, p. 4; AndreAnita/Getty Images, p. 5; Gabrielle Therin-Weise/Getty Images, p. 6; Thomas Kokta/Getty Images, p. 7; Staffan Widstrand/ Getty Images, p. 8; Patrick J. Endres/Getty Images, p. 9; xiaoying shi/Getty Images, p. 10; www.hepherson.com/Getty Images, p. 11; Paul Souders/Getty Images, pp. 12, 17, 18; Galaxiid/ Getty Images, p. 13; Jeff Foott/Getty Images, p. 14; GTW/Shutterstock, p. 15; VargaJones/ Getty Images, p. 16.

Cover credit: Keren Su/Getty Images.